Taking Care of Myself

Getting Sleep

By Meg Gaertner

level 2
little blue readers

www.littlebluehousebooks.com

Little Blue House is distributed by North Star Editions:
sales@northstareditions.com | 888-417-0195

Produced for Little Blue House by Red Line Editorial.

Photographs ©: Shutterstock Images, cover, 4, 7, 8, 11, 12, 17, 19, 20–21, 23, 24 (top right), 24 (bottom left); iStockphoto, 15, 24 (top left), 24 (bottom right)

Library of Congress Control Number: 2021916805

ISBN
978-1-64619-492-6 (hardcover)
978-1-64619-519-0 (paperback)
978-1-64619-570-1 (ebook pdf)
978-1-64619-546-6 (hosted ebook)

Printed in the United States of America
Mankato, MN
012022

About the Author

Meg Gaertner enjoys reading, writing, dancing, and being outside. She lives in Minnesota.

Table of Contents

Getting Sleep 5

Not Enough 9

Better Sleep 13

Glossary 24

Index 24

Getting Sleep

Sleep gives the body a break. It helps the body prepare for tomorrow.

Everybody needs sleep. Kids need 9 to 12 hours of sleep each night.

Not Enough

Sometimes we don't get enough sleep.
Then we feel tired.

We can’t think as quickly or clearly.

We get mad more easily.

Certain actions are harder than usual.

exercise

Better Sleep

You can get better sleep. Exercise during the day. Being active helps people sleep better.

Do not have sodas
or sugary drinks
near bedtime.
These drinks make people
stay awake.

sodas

Turn off the TV one hour before bedtime.
Turn off phones and computers, too.

TV

Do something quiet and peaceful before bedtime. You could take a warm bath or read a good book.

book

Go to bed at the same time each night. Your body will get used to falling asleep then.

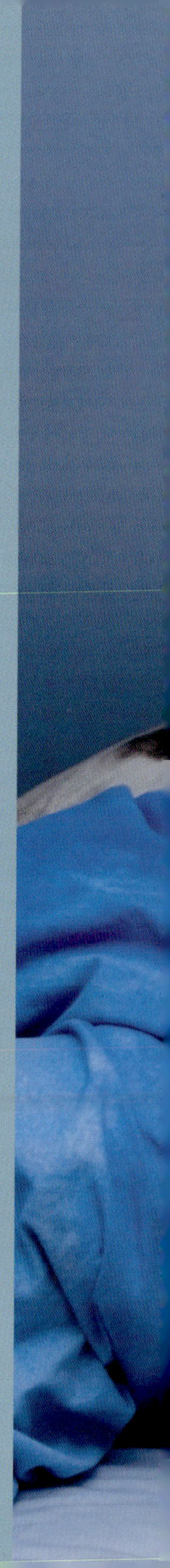

bed

These tips will help you get better sleep. You will feel rested in the morning. You will be ready to have a great day.

Glossary

bath

exercise

computer

sodas

Index

B

bath, 18

D

drinks, 14

E

exercise, 13

P

phones, 16